Bibliographic information published by the German National Library:

The German National Library lists this publication in the National Bibliography;
detailed bibliographic data are available on the Internet at http://dnb.dnb.de .

Imprint:

Copyright © 2018 GRIN Verlag
Print and binding: Books on Demand GmbH, Norderstedt Germany
ISBN: 9783668861305

This book at GRIN:

https://www.grin.com/document/451366

Christian Lienen

The usage of FPGAs for the acceleration of Convolutional Neuronal Nets (CNNs) with OpenCL. Two alternatives for implementation

GRIN Verlag

FPGA-based CNN Accelerators with OpenCL

Christian Lienen
Paderborn University

Abstract

Convolutional Neuronal Nets (CNNs) are state-of-the art Neuronal Networks which are used in many fields like video analysis, face detection or image classification. Due to high requirements regarding computational resources and memory bandwidth, CNNs are mainly executed on special accelerator hardware which is more powerful and energy efficient than general purpose processors. This paper will give an overview of the usage of FPGAs for accelerating of computation intensive CNNs with OpenCL. Therefore, two different implementation alternatives are proposed. The first approach is based on nested loops, which are inspired by the mathematical formula of multidimensional convolutions. The second strategy transforms the computational problem into a matrix multiplication problem on the fly. The approaches are followed by common optimization techniques used for FPGA designs based on high level synthesis (HLS). Afterwards, the proposed implementations are compared to an CNN implementation on an Intel Xeon CPU. This makes it possible to demonstrate the advantages in terms of performance and energy efficiency.

1   Introduction

Convolutional Neuronal Networks (CNNs) are a class of feed-forward artificial neural networks which are inspired by the visual cortex of the brain. They are commonly used for computer vision applications like face recognition, image classification or a wide range of speech recognition applications. During the last decade, there were significant accuracy improvements due to enhanced network structures. Some of these improvements are made possible by advances in the field of integrated circuits, which enable the processing of large amounts of data through reduced structure sizes and massive parallelism. Besides that, embedded application such as Google's reverse image search are running on devices with limited computational power and small battery lifetimes. This is another reason why energy-efficient implementations are needed. The training is usually done on high performance GPUs and other accelerators like Intel Xeon Phi coprocessors [2]. On the other hand some vendors create their own specific accelorator processors, like Google's Tensor Processing Unit [4].

Field programmable gate arrays (FPGA) have been increasingly used for this purpose for several years. The advantages using FPGAs are flexibility and energy efficiency. Unlike ASICs, FPGAs can be reprogrammed in the field. Since they have thousands of computational units and fast on-chip memory make itself useful for massive parallel computations. Due to the usage of high level languages like OpenCL for the development, the time to market could be reduced. Thus, also larger designs can be realized in an acceptable time.

This paper is structured in the following way. Chapter 2 presents related work for this research field. In chapter 3, background information regarding FPGA technology and deep neuronal networks with CNN layers is presented. Two different approaches for implementing CNNs on FPGAs are shown in chapter 4. Since optimization is a big challenge for designing FPGA accelerators, common techniques are described there. Before concluding, a comparison between the two implementations and a CPU implementation follows.

2 Related Work

Several accelerators for computation intensive problems have been proposed over the last decade. Most of the computational effort of CNNs is effected by the convolution operations which can basically be implemented sequentially by nested loops or by matrix multiplications calculus. Matrix multiplications are used by Zhang et al. ([9]) and Suda et al. ([7]). The nested loop approach is used by Zhang ([8]) as a base for further optimization.

C. Zhang et al. ([8]) and C. Zhang et al. ([10]) consider of on-chip buffers for data reuse technique in order to optimize the CNN computation and memory access. Much research has been done in optimizing data representation (Gupta et al., [3]; Courbariaux et al., [1]). Another approach for data type optimization is introduced by Z. Zhang et al. ([10]). They reduced the floating-point data size to 9 bits and implemented quad multiplier operation on a single DSP element.

3 Background

While GPUs and other specialized hardware like Intel Xeon Phi processors are widely used for accelerating computing systems, FPGAs have been integrated in high performance computing environments recently. For example cloud platform providers, such as Amazon and Microsoft, offer FPGAs to their customers. The use of OpenCL as a hardware description language contributes to this significantly.

3.1 FPGA

Field Programmable Gate Arrays are standardized integrated circuits whose functionality can be configured after production in the field. On the chip, there are standardized digital elements. The configuration and the connection between them are specified in the synthesis process.

The main components of the chip are logic elements which structure depends on the specific FPGA type. In general, logic elements consists of one or more lookup tables which have a register on it's output. The lookup table is needed to realize combinatoric logic functions (e.g logical and or logical or-functions). The registers are used to store the result for one or more cycles.

Additionally, FPGAs consist of configurable components like on-chip memory, routing networks, clock-trees and digital signal processing (DSP)-elements which allow higher clock frequencies than similar computational units built with logic elements. On-chip block ram memory (BRAM) can be used for full performance access of data for the computational units. Most of these blocks have two independent memory ports which allow independent access (read or write) operations at the same time.

For accelerator purposes, FPGAs are placed on PCI express accelerator cards. In addition to the FPGA chip on the board, the accelerator cards have components for power supply, external memory and other interfaces, such as PCIe or Ethernet.

The design implementation for large FPGAs is meanwhile done in high level synthesis (HLS) languages like SystemC or OpenCL. Thus, the time to market is significant reduced compared to classical hardware description languages like VHDL or Verilog. The global market is dominated by two big vendors: Intel (previously Altera) and Xilinx. Both are providing individual toolkits for the development of FPGA designs to their costumers.

3.2 CNNs

Convolutional Neuronal Networks (CNN) are a class of feed-forward artificial neuronal networks which are inspired by the human cortex. State-of-the-art CNNs consist of one or more convolution layers combined with non-linear activation functions, normalization functions and pooling layers. Normalization functions are intended to normalize each output by a factor which depends on the neighbors of that specific neuron. The normalization yields to a standard normal distribution of the output data. Non-linear activation functions like ReLu $f(x) = max(0, x)$ or Sigmoid $g(x) = (1 + \exp(-t))^{-1}$ are added to decouple layers from each other and to introduce non-linearity into the net. Pooling layers are used to reduce the feature dimensions. The last layers of the whole net are usually built as fully-connected layers for classification. Fully connected layers are neuronal nets where all inputs of the next layer are connected to each output of the previous layer.

The most critical part regarding computational performance in CNNs are the convolution operations which contributes most of the computational effort to the neuronal net. They create a set of feature maps from the input by compute convolutions on the input feature map. As shown in figure 1, the input feature map (left side) is transformed into one 2D output map (right side) with two $K \times K$ filter kernels (middle).

In general, convolution layers generate N_{out} output feature maps from N_{in} input feature maps. Every dimension of the output feature map is the result of the convolution with a $K_h \times K_w \times N_{in}$ filter kernel, where K_h is the height and K_w is the width of the filter ker-

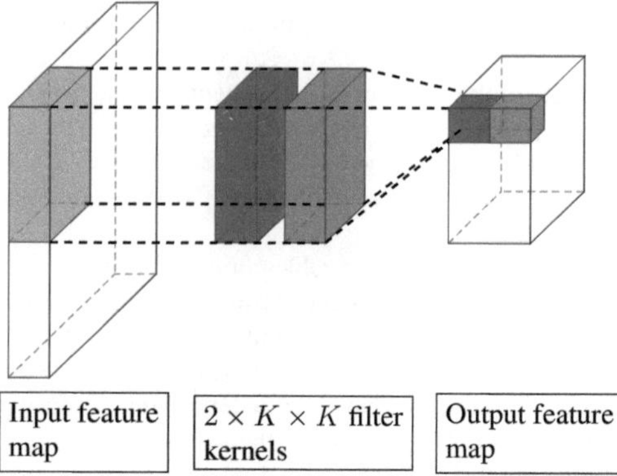

Figure 1: CNN operation: The scalar product of input times filter results to the output feature map element

nel. Typically, the kernel is squared such that $K_h = K_w = K$. In fact, there exists N_{out} of these kernels such that the overall dimension of the filter kernel is $K_h \times K_w \times N_{in} \times N_{out}$. The computation of element at position (x_w, x_h) in the i-th output map from the j-th input map is shown in equation 1. References to negative indexes or to indexes outside of the input map are responded with zero. For computing the complete set of output feature maps, the number of sums increase from two (equation 1) to six.

$$Y_i(x_w, x_h) = \sum_{k_w=0}^{K_w} \sum_{k_h=0}^{K_h} W_{i,j}(k_w, k_h) \cdot X_k(x_w - k_w, x_h - k_h) \tag{1}$$

4 Implementation

The implementation of CNNs consumes a considerable amount of computational resources as well as also a significant amount of memory. Large CNN-based architectures like AlexNet have more than 60 million model parameter. These parameters consume about 250 MB memory assuming 32 bit data type size. Since FPGAs do not provide on-chip memory on this scale, these parameters have to be stored in external memory. Therefore, the external memory bandwidth can become a performance bottleneck. Because of that, the challenge in the implementation part is to optimize the flow of data to and from the compute units.

In this chapter, two different approaches for CNN computation are presented. The first approach uses nested loops for the implementation which corresponds to the implementation of equation 1 basically. The second approach transforms the problem into matrix multiplications on-the-fly. In this form, standard matrix multiplication architectures can be used for computation.

4.1 OpenCL Stack on FPGA

Previously, FPGA designs were described by hardware description languages like VHDL and Verilog. For some time, the large FPGA vendors offer high-level language compilers, which enable faster hardware development compared to these classical hardware description languages. Besides shorter time-to-market cycles, the kernel code can be easily migrated to other platforms like CPUs or GPUs .

The specification of the FPGA design is done by the standard OpenCL programming model. This standard model is extended by some vendor specific functions like interfaces for kernel-to-kernel communication or parameters to influence the synthesis process (pragma instructions). The resulting code is compiled by the OpenCL toolkit to the FPGA configuration binary file. To this end, the OpenCL code is compiled to an intermediate description language at first. At this time, infrastructure components are also added. This includes the PCIe-Interface for communication to the host CPU, the memory controller for the external DRAM or infrastructure for kernel-to-kernel and kernel-to-host communication. In the next step, the synthesis process transforms the intermediate result into the FPGA design. Unlike other fixed architectures like GPUs, this process has to be done offline. This means that the binary file has to exist before program executing. The final design is transmitted to the FPGA during host code execution time (figure 2).

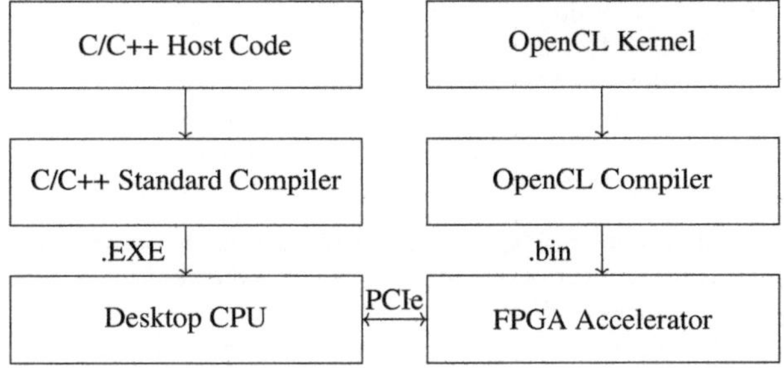

Figure 2: OpenCL-based Design flow at Host and Kernel [7]

The programming model of OpenCL allows at least two different modes of kernel operation: Single-work-item execution and Multi-work-item execution. Single-work-item kernels are instantiated one time and executed until the program is elapsed. The whole computational problem is computed by this instance. For the Multi-work-item approach, one or more instances of the execution units are instanced. The work-items are allocated to these execution units by the dispatcher. The work-items are organized into work-groups. Unlike the Single-work-item approach, instances of the execution unit can execute multiple work items sequentially. The kernel region for the multi-work-item approach consists of three basic parts: a dispatcher, a compute subsystem which is organized into compute units and processing elements, and a memory subsystem. The dispatcher allocates work-items to the compute units and controls host-to-kernel and device memory communication. Each compute unit has its exclusive local memory and access to the shared global memory.

On the host side, C/C++ code is typically used, which is translated into machine code using a standard compiler. While executing, the host program loads the binary file for the FPGA. After loading, the host program initiates the memory transfer to the FPGA accelerator and starts the kernel execution.

4.2   Implementation as Nested Loops

The nested-loop approach which computes similar to equation 1 is proposed by Zhang in 2015 [8]. The computation of all output feature maps is done by at least six nested loops. The basic loop structure is shown in the Algorithm 1.

Algorithm 1 Nested loop convolution

---

for $row \leftarrow 1, R$ **do**
    for $col \leftarrow 1, C$ **do**
        for $to \leftarrow 1, M$ **do**
            $O(to, row, col) \leftarrow 0$
            for $ti \leftarrow 1, N$ **do**
                for $i \leftarrow 1, K$ **do**
                    for $j \leftarrow 1, K$ **do**
                        $p \leftarrow weights(to, ti, i, j) \cdot I(ti, row + i, col + j)$
                        $O(to, row, col) \leftarrow O(to, row, col) + p$

The first two loops are running over the width and height dimension of the input feature map. M is the number of output maps and N is the number of input maps. The filter kernel has the dimension $K \times K$.

This approach generates several design challenges which have to be solved. First, all data elements are stored into external memory. For efficient caching of the data in the on-chip memory, loop tiling is necessary to fit these memory constraints temporarily. On the other hand, data dependencies, specially for the variable p in the algorithm, prevent efficient loop pipelining. Thus, additional optimization techniques like distributed accumulations have been applied.

The overall implementation of the design is shown in figure 3. The presented loops are implemented in the compute engines, which are shown in the middle of the picture. For accelerating of the external memory access, four buffer banks are added, which are connected to the compute units and to the external memory controller. The external controller controls the data exchange between the buffer banks and the external memory. Additionally, it can isolate the design from various platforms and vendor specific properties. Both input buffer banks are used for the data of the input feature maps and weights. The other two memory banks are used for the resulting output feature maps. The buffers are also

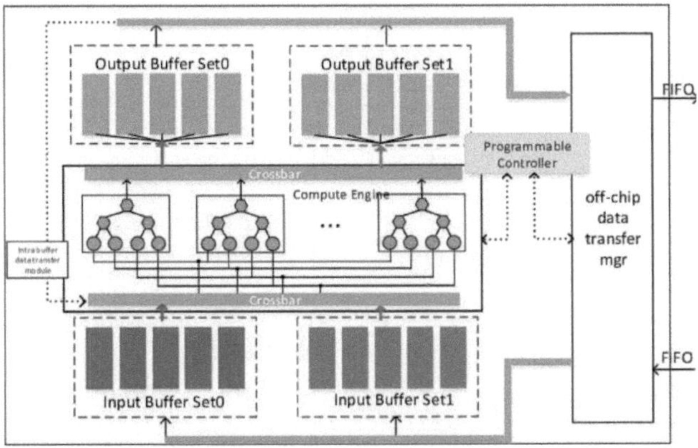

Figure 3: OpenCL compute architecture [8]

used to overlay computational time with transfer time. The programmable controller from
the picture works as control element.

4.3 Implementation as Matrix Multiplication

Zhang et al. [9] and Suda et al. [7] describe an alternative approach. The calculations of
the convolution operations are done by a matrix multiplications $A \cdot B = C$. The kernel
weights are ordered row-wise in the matrix $A(M \times N)$ for the multiplication. The input
features of the convolution kernel are placed column-wise in $B(N \times P)$. The ordering
is shown in Figure 4. By definition, every element $C(M \times P)$ of the resulting matrix is
the scalar product of row n of A times column m of B. This is equal to one convolution
filter operation over all input feature maps (equation 1). The set of all partial results then
results in a matrix again. The input features are rearranged on-the-fly while moving from
the external data memory to the internal on-chip memory.

The whole matrix multiplication is divided into several submatrix-multiplications to
enable parallelization of the computation. These sub-problems are matrix multiplications
of type $(N_{conv} \times N) \times (N \times N_{conv}) \rightarrow N_{conv} \times N_{conv}$. Therefore, the input feature
dimension P and the related kernel weights dimension M have to be a multiple of a
factor N_{conv}, which can also be reached by zero padding. M and N are the dimensions of
original matrices. Every sub-problem is calculated in a separate work-item. The proposed
divide and conquer strategy allows parallelism on a high level. Besides that, the division
allows also more efficient data caching.

The computation of this approach is done in the computational subsystem shown in
figure 5. The OpenCL framework on the top-level is divided into the infrastructure region

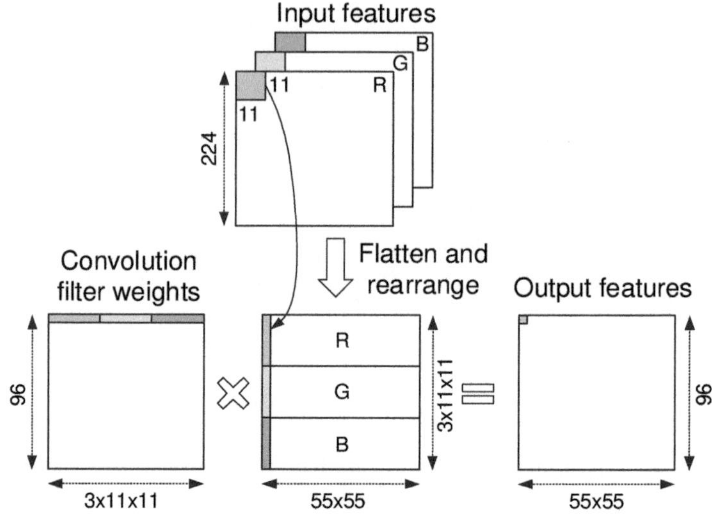

Figure 4: Conversation of a convolution into matrix multiplication [7]

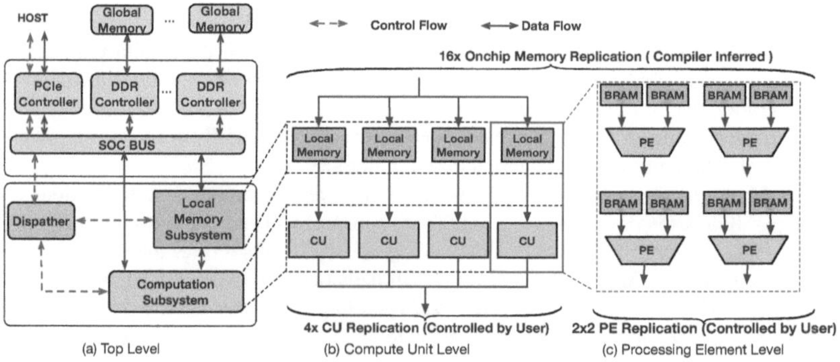

Figure 5: OpenCL Matrix Multiplication Architecture [9]

and the kernel region. The kernel space of the architecture consists of a local memory subsystem, a computational subsystem and a dispatcher. The computational resources are replicated on two different levels. On the higher level, the computational subsystem and the local memory subsystem is divided into four compute unit replications. Every of these replications is divided into four processing elements again. The processing element are the on lowest level of the computational system and executes one MAC-operations (Multiply-and-Accumulate) per cycle. The number of replications is controlled by the user and depends on the available resources of the FPGA. Each of these processing elements are supplied by two BRAM ports. Besides that, all compute units share a global memory and a constant memory.

The dispatcher is responsible for the work-item schedule and the assigning of work-items to related compute unit elements. Furthermore, the dispatcher also controls host-device memory transfers. The infrastructure region consists of the PCIe-engine which also includes a DMA-controller (DMA: Direct memory access). Additional there are one or more external memory controllers.

5 Optimizations Techniques

The following section focuses on optimization techniques for OpenCL FPGA designs. In general, there are two different main approaches in this field: computational optimization and data path optimization. Computational optimization is applied to increase the number of operations by the design through data level parallelism like improved pipelining, vectorization or loop unrolling. On the other hand, data path optimization also increase the throughput of the processing elements. This approach focuses on the data path between external memory and the compute units in the FPGA. This kind of optimization helps to increase the utilization of the DSP elements during synthesis and during runtime. For an efficient usage of either computational or data path optimization, understanding of the performance limiting factors is needed. To do so, performance modeling is applied to the design.

An alternative to the manual application of the presented optimization is presented by Pouchet et. al. [5]. They present a fully automated C-to-FPGA framework to address optimization problems. The focus is on the improvement of the execution time of loops. This is done by effective utilization of on-chip memory. This goal is reached by aggressive loop transformations and restructuring which enable effectively data reuse. Moreover, critical optimization like task-level parallelization, loop pipelining and data prefetching are inserted by the framework.

5.1 Computational Optimizations

In the following chapter, properties of loop-based execution are introduced. The number of cycles, which are needed for one full iteration of a loop starting with the first operation of the body and ending with the last operation is called Latency L. For non-

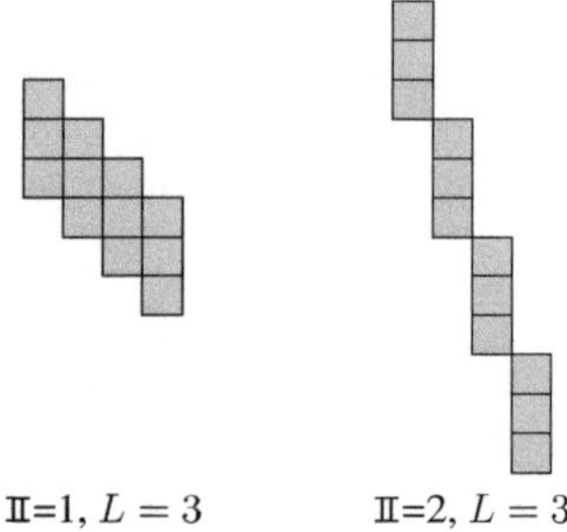

Figure 6: Different loop computation schedules

optimized loops with M iterations (without any parallelism) the needed number of cycles is $N_{cycles} = L \cdot M$. This case is shown in the right part of figure 6. Every square symbolize one single operation of the loop body.

In the following sections techniques for parallelism are presented, which allow overlapping executions from different loop iterations. This case is shown in the left part of the figure 6. The first atomic operation of the following iteration can start directly after finishing the operation of the current iteration. The resulting initiation interval II for this computational schedule is II $= 1$. The initiation interval is the number of cycles before the next iteration can start. Due to overlapping execution, the execution time for the whole loop is $N_{cycles} = $ II $\cdot M + L$. The latency of the body has to be paid one time in the beginning. After that, the execution unit delivers one output value per cycle assuming II $= 1$.

5.1.1 Improved Pipelining

Loop pipelining is the standard technique of high level synthesis to increase the computational performance. The performance optimization is reached by overlapping execution from different loop iterations. Loop pipelining is limited by data dependencies and available hardware resources. The data dependencies are also called loop-carried dependencies and imply that the following iteration of the loop can not start until the actual loop iteration is done. Therefore, the minimum initiation interval is limiting this kind of dependencies. For perfect pipelining, loop-carried dependencies have to be avoided because they yields so called bubbles in the pipeline and therefore performance losses. The theoretical speedup S_k by using pipelining is also affected by the latency for pipeline filling at the beginning. The effect dominates especially for small loop bounds. The overall speedup is shown in equation 2, where n is the number of operations and k the number of stages. For $n \to \infty$ the speedup is k.

$$S_k = \frac{n \cdot k}{k + n - 1} \tag{2}$$

One approach to solve loop-carried dependencies for MAC operations is to use distributed accumulations in the loop. The accumulator is divided into p different variables and summarized afterwards. The value p depends on the length of the MAC pipeline generated

```
for(int i=0; i<N; i++)
  accu[i%p] += w[i]*m[i];
```

by the synthesis. p should be a power of two to reduce the strength of the operation $i \bmod p$.

5.1.2 Loop-Unrolling

Another form of data level parallelism is the usage of loop-unrolling. This can be done manually or automatically by the compiler. Loop unrolling creates multiple copies of the loop body and execute them in parallel. Basically an unrolling-factor of N_u creates N_u instances of the loop body. The loop iteration counter is updated accordingly (e.g. $i+=N_u$). The resulting operation schedule is shown in figure 7. While the latency and the initiation interval are still the same, the overall calculation time is reduced.

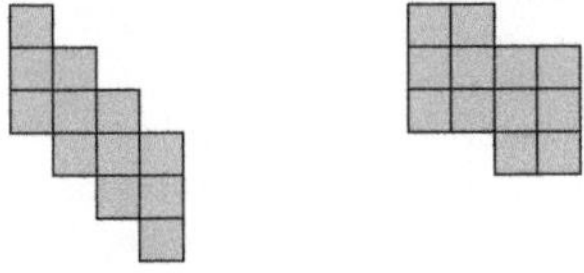

Figure 7: Loop schedule without (left) and with loop unrolling (right)

Loop unrolling is also limited by data dependencies between loop iterations and the number of available hardware resources (e.g. DSP elements or memory banks). The data exchange between two unrolled execution instances will affect the complexity of the generated control hardware and also the number of unrolled instances. The relation of the shared data between different instances can be divided into three classes:

- *Irrelevant* If the loop iterator variable j_k does not access array A, the loop dimension if j_k is irrelevant to that array A.

- *Independent* If the data space of an array A is totally separable along a loop dimension j_k, the loop dimension is independent of array A. In other words, the data space of a loop dimension j_k is disjoint to any other loop iterator $j_l, l \neq k$.

- *Dependent* If the part accessed by the loop level iterator j_k is not fully separable, the loop dimension j_k is dependent of array A.

Since memory areas are fully separable, an independent data sharing relation yields to direct connections between buffers and computation engines. In this term of loop unrolling, computation engines are loop body instances. Irrelevant relations generate are direct connection to any processing elements. On these connections, the data is distributed via broadcast. However, data dependent relations led to complex data distribution structures between memory and processing elements.

5.1.3 Vectorization

The goal of using vectorization is to reduce the overhead due to control structures. This is because of multiple data elements can be processed at the same time. The same principle is also used in modern CPUs using SIMD-instructions (Single instruction multiple data). OpenCL supports either fixed-point and floating-point vector data types with a power-of-two number of elements up to 16.

5.2 Datapath Optimizations

One of the most critical challenges in developing FPGA accelerator designs is the optimization of the data flow between external memory and the computational units. Due to improvements in last decades, the computational performance of FPGAs has increased significantly, but the on-chip memory size and performance did not grow in that scale. Therefore, the on-chip memory has to be used in an efficient manner.

One common technique for data flow optimization is data reusage. The idea behind this technique is to use data for multiple computational operations after loading from external memory. As a consequence, the relative overhead regarding external data access decreases.

Zhang and Li [9] introduced the metric of machine balance B_m to quantify the ratio between memory bandwidth and computational bandwidth of the system. B_m is defined as the ratio of the maximum achievable memory bandwidth to peak arithmetic performance provided by the hardware. The metric B_m^{on} is defined as the machine balance for internal memory and can be calculated as follows:

$$B_m^{on} = \frac{N_{BRAM} \cdot f_{DSP}}{N_{DSP} \cdot f_{BRAM}} \tag{3}$$

N_{BRAM} is the number of BRAM with working frequency f_{BRAM}, N_{DSP} the number of DSP elements with the working frequency f_{DSP}. The reason for a low B_m^{on} and therefore computational losses through low memory bandwidth can be found in the unnecessary replications of memory during processing element replication. An increasing efficiency of BRAM usage yields to higher computational performance.

For the matrix multiplication approach, the data access pattern create opportunities for data reusage by multiple processing elements for one on-chip BRAM port. The following equation 4 shows the matrix multiplication scheme.

$$(\boldsymbol{AB})_{ij} = \sum_k \boldsymbol{A}_{ik}\boldsymbol{B}_{kj} \tag{4}$$

For calculating $\boldsymbol{AB}$, the scalar product of each row vector of $\boldsymbol{A}$ times each column vector of $\boldsymbol{B}$ has to be calculated. Therefore, multicast connections between BRAM and processing elements can be created such that not only on-chip bandwidth without replication increases but also on-chip data reuse improves by $\sqrt{N_{DSP}}$. The BRAM requirement for N_{DSP} processing elements is reduced from $\mathcal{O}(N_{DSP})$ to $\mathcal{O}(\sqrt{N_{DSP}})$.

$$B_m^{on} = \sqrt{N_{DSP}} \cdot \frac{N_{BRAM} \cdot f_{DSP}}{N_{DSP} \cdot f_{BRAM}} = \sqrt{N_{DSP}} \cdot B_m^{on} \tag{5}$$

The implementation of multicast interconnection between BRAM and processing elements requires modifications to the dispatcher for the work-items. The current OpenCL dispatcher schedules the work-items along the lower-dimension. The scheduler has to process block-wise for enabling multicast interconnections , which can be realized by work-groups. The optimal size of the work-groups is a separate optimization problem ([9]).

6 Comparison

This section presents the implementation results of both proposed approaches. The optimization techniques, which are described in the previous chapter, are applied to the designs. In general, FPGA designs which are implemented on FPGA from different vendors and architectures are not fully comparable. There are many aspects like e.g. the technology and structure sizes, the structure of logic and the structure of DSP elements, which prevent a fair comparision. For all comparisions of experimental results in this chapter, the VGG net (Very Deep Convolutional Networks for Large-Scale Image Recognition) [6] is used. The compilation and synthesis of the matrix multiplication approach is done with Altera OpenCL SDK. The Xilinx Vivado SDK is used for the nested-loop approach. Besides that, the outputs of the tool chains like the utilization of the FPGA the performance is given as giga operations per second. Because of the different frequencies of the systems and the different FPGA types, the metric of performance density is introduced. This metric shows the relative performance of the implementation to the available DSP elements and the reached clock frequency. Other parameters like the concrete structure of the DSP elements are not considered by this metric.

Device	Matrix Multiplication Impl.	NestedLoop Impl.
Device	Arria 10	Virtex 7
Toolchain	Intel Quartus	Vivado toolset
Frequency (MHz)	370 MHz	100 MHz
DSP Utilization	1320/1518	2240/2800
BRAM Utilization	1250/2713	1024/2060
Precision	float	float
Performance (Gops/s)	866	61.62
Perfomance Density (ops/DSP/cycle)	1.54	0.22
Power Efficiency (Gops/s/W)	20.75	3.31

Table 1: Comparison between both FPGA implementations

To classify the results in a larger context, a comparison between the implementations and a CPU-based implementation of VGG is shown in Table 2. For testing, the OpenCL nested-loop approach is compiled for a CPU. The software counterpart of the FPGA designs are running with 16 threads. The power consumption is assumed to be 95 Watt which is the TDP (TDP: thermal design power) property of the processor. The real power consumption may be lower than the TDP but it is sufficient to show the order of magnitude. The comparison shows the performance improvements and the massive gain in energy efficiency of up to 153 times.

Device	CPU Xeon	Virtex 7	Arria 10
Frequency	2.2 GHz	100 MHz	370 MHz
Precision	float	float	float
Performance (Gops/s)	12.87	61.62	866
Speedup	1.0	4.79	67.29
Power (W)	95.0	18.61	41.74
Power Efficiency (Gops/s/W)	0.135	3.31	20.75
Gain in efficiency	1.0	24.59	153.70

Table 2: Comparison between CPU and FPGA implementation

7 Conclusion

State-of-the-art Convolutional Neuronal Nets a widely accepted for solving vision and speech recognition challenges. Many applications require either high performance but also energy efficiency because of limited batteries. Modern FPGA technology enables fast and energy efficient accelerator designs for Convolutional Neuronal Nets. By introducing OpenCL for FPGA, the gap design description between developer efficiency and technology progress is reduced. For accelerating deep neuronal networks based on convolution operations, two different approaches are shown. The first approach is based on nested loops, which are inspired by the mathematical formula of multidimensional convolutions. The second strategy transforms the computational problem on-the-fly into a matrix multiplication problem. After implementation, common optimization techniques are described. This topic is divided into a computational and a data flow optimization part. The computational part describes optimized pipelining, loop unrolling and vectorization. The datapath optimization part introduce the metric of memory balance, which helps to analyze the data flow from computational units and external memory. Due to the reusage of data elements, the on-chip memory usage can be reduced. The last chapter shows a comparison between the presented architectures and an implementation of the nested-loop approach on a common CPU. Although comparisons between designs of different FPGA types are not fully comparable, the matrix multiplication approach enables better performance compared to the nested-loop approach. On the other hand, both approaches are better than the example CPU implementation by orders of magnitude. All in all, the proposed implementations yield better energy efficiency and performance compared to CPU-based implementations, although the presented topic is still an active research topic.

References

[1] Matthieu Courbariaux and Yoshua Bengio. Binarynet: Training deep neural networks with weights and activations constrained to +1 or -1. *CoRR*, abs/1602.02830, 2016.

[2] Matthew Dixon, Diego Klabjan, and Jin Hoon Bang. Implementing deep neural networks for financial market prediction on the intel xeon phi. In *Proceedings of the 8th Workshop on High Performance Computational Finance*, WHPCF '15, pages 6:1–6:6, New York, NY, USA, 2015. ACM.

[3] Suyog Gupta, Ankur Agrawal, Kailash Gopalakrishnan, and Pritish Narayanan. Deep learning with limited numerical precision. *CoRR*, abs/1502.02551, 2015.

[4] Norman P. Jouppi and Cliff Young et al. In-datacenter performance analysis of a tensor processing unit. *CoRR*, abs/1704.04760, 2017.

[5] Louis-Noel Pouchet, Peng Zhang, P. Sadayappan, and Jason Cong. Polyhedral-based data reuse optimization for configurable computing. In *Proceedings of the ACM/SIGDA International Symposium on Field Programmable Gate Arrays*, FPGA '13, pages 29–38, New York, NY, USA, 2013. ACM.

[6] Karen Simonyan and Andrew Zisserman. Very deep convolutional networks for large-scale image recognition. *CoRR*, abs/1409.1556, 2014.

[7] Naveen Suda, Vikas Chandra, Ganesh Dasika, Abinash Mohanty, Yufei Ma, Sarma Vrudhula, Jae-sun Seo, and Yu Cao. Throughput-optimized opencl-based fpga accelerator for large-scale convolutional neural networks. In *Proceedings of the 2016 ACM/SIGDA International Symposium on Field-Programmable Gate Arrays*, FPGA '16, pages 16–25, New York, NY, USA, 2016. ACM.

[8] Chen Zhang, Peng Li, Guangyu Sun, Yijin Guan, Bingjun Xiao, and Jason Cong. Optimizing fpga-based accelerator design for deep convolutional neural networks. In *Proceedings of the 2015 ACM/SIGDA International Symposium on Field-Programmable Gate Arrays*, FPGA '15, pages 161–170, New York, NY, USA, 2015. ACM.

[9] Jialiang Zhang and Jing Li. Improving the performance of opencl-based fpga accelerator for convolutional neural network. In *Proceedings of the 2017 ACM/SIGDA International Symposium on Field-Programmable Gate Arrays*, FPGA '17, pages 25–34, New York, NY, USA, 2017. ACM.

[10] Z. Zhang, D. Zhou, S. Wang, and S. Kimura. Quad-multiplier packing based on customized floating point for convolutional neural networks on fpga. In *2018 23rd Asia and South Pacific Design Automation Conference (ASP-DAC)*, pages 184–189, Jan 2018.

YOUR KNOWLEDGE HAS VALUE

- We will publish your bachelor's and master's thesis, essays and papers

- Your own eBook and book - sold worldwide in all relevant shops

- Earn money with each sale

Upload your text at www.GRIN.com and publish for free